Henry Mason

Lectures upon the heart, lungs, pericardium, pleura, aspera arteria, membrana intersepiens or mediastinum

Henry Mason

Lectures upon the heart, lungs, pericardium, pleura, aspera arteria, membrana intersepiens or mediastinum

ISBN/EAN: 9783337273361

Printed in Europe, USA, Canada, Australia, Japan

Cover: Foto ©berggeist007 / pixelio.de

More available books at **www.hansebooks.com**

LECTURES,

UPON THE

HEART,	ASPERA ARTERIA,
LUNGS,	MEMBRANA INTER-
PERICARDIUM,	SEPIENS, or ME-
PLEURA,	DIASTINUM.

Together with the DIAPHRAGM,

INTERSPERSED WITH

A Variety of PRACTICAL REMARKS.

By H. MASON, Surgeon.

Nulla enim re magis, quam exemplo docemur.

The above LECTURES were lately delivered at
the SURGEONS THEATRE.

R E A D I N G:
Printed for J. CARNAN, and Co. in the *Market-Place*; and fold by J. NEWBERY, at the *Bible* and *Sun*,
in St. Paul's Church-Yard, London.

TO

Mr. GRINDALL,

Senior Surgeon to the *London Hospital*,

A

Member of the Company of *Surgeons* of *London*,

AND

Fellow of the *Royal Society*.

SIR,

I Am induced to address the following LECTURES to you, from the Knowledge I have some Years had of your Abilities in Surgery; as also, from the favourable Cha-

A 2 racter,

racter, you was pleafed to inform me, the late Mr. GIRLE and Mr. NOURSE gave you of them, when they were read at the SURGEONS THEATRE.

I am, with the greateft Sincerity and Refpect,

Your moft obedient,

Humble Servant,

HENRY MASON.

The Introduction.

ANATOMY, when reftricted to
Surgery, is that art, which teaches
the fituation, figure, connexions,
fabrick, actions, and ufes of the feveral parts of the human body.

The intent and ends of this fcience are various: an admiration of one of the nobleft
works of the creator: a prefervation of our
health, as nothing can lead us more immediately to a knowledge of the means of preferv-
ing it, or reftoring it, when impaired by difeafes, than a true knowledge of the ftructure
of that frame which is injured by them.

Can a furgeon, who is unacquainted with the
human frame, make, for inftance, a deep incifion into any part of the human body, without
running a rifque of opening fome confiderable
artery, or wounding fome nerve?

Is

Is it not a manifeſt expoſing of the patient to terrible accidents, to bleed him, without knowing the parts adjacent to the vein we are about to open?

Unleſs he is well acquainted with the ſituation of the parts from anatomy, he will be always trembling, either with a vain fear, or elſe with a raſh aſſurance he will deſpiſe the danger of which he is ignorant.

Every chirurgical operation, therefore, proves how indiſpenſably neceſſary the knowledge of the parts of the human body is to Surgeons; and that flouriſhing ſtate of ſurgery we now ſee it in, is principally owing to thoſe beautiful diſcoveries with which anatomy has been lately enriched.

The want of this *index magneticus* (for as the needle is to a mariner, that is anatomy to a Surgeon) will in no inſtances be more felt by a Surgeon, than in giving his teſtimony in a court of judicature upon the untimely death of perſons he ſhall be appointed to examine; and particularly in caſes of wounds, for if the part is found to have loſt its uſe after the wound is healed, and as the judge uſually inflicts a penalty proportionable to the damage the patient ſuſtains, it is for this reaſon the council generally

rally ufe all the art they are mafters of to throw moft of the ill confequences upon the neglect or mifmanagement of the furgeon; therefore the ill effects which naturally follow fome wounds, though ever fo fkillfully treated, fhould be declared very early, and which can be deduced only from a knowledge of the anatomy, and injured functions of the parts wounded.

. But one who is acquainted from anatomy or phyfiology with the ufes of the parts, as far as they are at prefent known, will determine the confequences or effects of the wound as foon as the parts affected are known.

A maid-fervant fell down with a glafs mug in her hand, and wounded her arm betwixt the carpus and cubitus; a profufe hæmorrhage alfo followed, from a divifion of the artery running under the flexor carpi ulnaris mufcle: the hæmorrhage was happily reftrained by compreffing the trunk of the artery againft the os humeri in the upper part of the arm; but then the patient complained of a numbnefs in her little finger, and in the middle of the next finger, which the furgeon judged to arife from the compreffure of the artery; but being confirmed in my opinion by the accurate tables of Euftachius, I boldly affirmed that the nerve was divided which goes to the little finger, and to the middle of the next adjacent finger, and that therefore this complaint was irremediable. The event demonſtrated the truth of my aſſertion: for after the cure was compleated, at my requeſt, ſhe frequently put her finger into the flame of a candle, without feeling the leaſt pain.——*Vid. Van Swieten Comment. in Aphor. Boerhavii. Vol.* i.

Another important end anatomy ſerves, is, the determining the cauſe of the death of perſons dying of natural diſeaſes, from a ſubſequent diſſection of the body——This may be called morbid anatomy.----Hereby we are ſhown the difference between the diſeaſed and healthy ſtate of the parts; and the latent cauſes of multitude of diſeaſes have been long ſince diſcovered by means of theſe diſſections, which otherwiſe we never could have known. Yet, there is here ſome caution neceſſary; for the dead body only ſhews us what its ſtate was at the time of death, and many changes will be found made in it by the diſeaſe, which, however they may be the effects of the diſeaſe, would be very improperly reckoned to be the cauſe of it.

The knowledge of the real cauſes of a diſeaſe, is the firſt rational ſtep towards its cure; and without the aſſiſtance of diſſections, how was the world to have known that the æſophogus
could

could have been burſt in a violent fit of vomiting, that the ſpleen growing to an immenſe ſize could have fallen into the pelvis by an elongation of its connecting veſſels, or have been informed of the nature of an empiema, a cataract, a hernia, and many others.

They are much miſtaken, ſays Baglivi, who think they can cure diſeaſes happily becauſe they are maſters of the Theory. They ought to have much higher things in view. They muſt diſſect the bodies of thoſe who die of diſtempers, and *foul their fingers*, to the end they may find out the ſeat of the complaint, the cauſe and iſſue of antecedent ſymptoms.

The ancients were ſo well convinced of the neceſſity of this knowledge, and to take, as they thought, the beſt method of attaining it, procured criminals out of priſon, and diſſecting them alive, contemplated while they were even breathing the parts which nature before had concealed, conſidering their poſition, colour, ſize, &c. for, ſay they, as various diſorders attack the internal parts, they thought no perſon could apply proper remedies to thoſe parts, which he was ignorant of.

The anatomical obſervations likewiſe made upon brutes have not only given great light to
thoſe

thofe made upon human bodies, but are exactly
of a piece with them, nay, fo certain and conftant is that mutual analogy, that the ftructure
of the vifcera, and an infinity of other things
have been difcovered to the great happinefs of
the age we live in.

Without thefe neither would Herophilus
have difcovered the lacteal veffels in kids; Euftachius and Pecquet their receptaculum chili
and thoracic duct in the horfe and dog, nor
Harvey his celebrated circulation of the blood.
Add to thefe that the infpections of carcafes by
priefts in their daily facrifices; the cuftom of
embalming and opening the dead, as alfo dreffing of carcafes by the butcher, each afforded
fome knowledge of the anatomical ftructure of
found bodies, as well as the immediate and
abftrufe caufes of health, ficknefs, and death.

All this could hardly be done for many ages
together, without frequently detecting the latent caufes of the moft fevere difeafes as well
as the ftructure and fituation of the parts;
and hence the firft foundation of practical anatomy.

Frequent and deftructive war furnifhed op-
portunies of difcovering many of the mufcles
and

and larger veffels, with the articulation of the bones, to the naked eye in the yet living fubject; infomuch that fome have attempted to extract a fyftem of anatomy from Homer, who has in reality writ hiftories of wounds fkilfully and anatomically ftated

We have been taught alfo from practical anatomy that difeafes frequently change the natural fituations of the vifcera, and this we are affured of from the moft certain obfervations. The pofition of the ftomach efpecially has been obferved to be furprizingly perverted together with the other vifcera of the abdomen, in the body of a woman after frequent vomitings. (*Mem. acad.* 1716, *page* 238.) And it feems very probable, that the vifcera are thus even frequently difplaced, fince, fays Van Swieten, I have feveral times made the like obfervation, in the fubjects which I either diffected myfelf, or have feen diffected by others. I have feen the fpleen prolapfed into the pelvis, the bottom of the ftomach continued below the navel; and have alfo feen that part of the colon which lies under the ftomach fo reflected thence as to form an arch below the navel, the convex part of which was towards the pelvis, and its concavity towards the ftomach.

The

The advantages which Surgery particularly derives from Anatomy being very evident, I fhall next proceed to what I intend in the following lectures, which is to confider thofe bowels placed in the cavity of the Thorax.

PRÆLECTIONES

PRÆLECTIONES

IN QUIBUS

Tractandæ veniunt partes quæ in
Thorace funt contentæ;

Unâ cum

Velamentis extus circumpofitis.

Diaphragma, quod hujus ventris
pars effe magis quam infimi,
hoc loco defcribetur.

Variis inde deductis corollariis practicis.

PRÆLECTIO PRIMA.

AS the Diaphragm contributes in great meafure to the formation of the cavity of the Thorax, and a knowledge of its fituation and ample extent being of fome confequence in practice, I fhall begin with a defcription of this mufcle.

We call the thorax that part of the trunk of the body which is terminated before by the fternum, behind by the twelve vertebræ of the back, on the fides by the arched ribs, above by the two fuperior ribs, and below by the diaphragm, which feparates it from the cavity of the abdomen.

The diaphragm, or the mufculus formâ mirus, as diftinguifhed by Albinus, is a very broad and thin mufcle, fituated at the bafis of the thorax, and which ferves to partition off, by a very broad furface, the lower vifcera, from thofe of the breaft. It is placed obliquely, and forms a kind of arched roof, or concave dome, with

its

its convex part towards the breaft, and in fuch a manner that its fore-part rifes much higher than its back-part, which is inferted lower, hence it is that the cavity of the thorax is much larger behind than before. Since the cavity of the thorax defcends deepeft towards the back, from the inclined pofture of the diaphragm, therefore when we are about to perform the paracentefis of the breaft, we fhould make our perforation as low as it can poffibly be done, without danger of injuring the diaphragm. And to avoid hurting the ftrong mufcles termed facrolumbalis, longiffimus, dorfi, &c. which afcend thro' the loins and back on each fide the fpina dorfi, the opening ought to be made at the diftance of four fingers breadth, at leaft, from the vertebræ, and this is ufually made betwixt the fecond and third, or betwixt the third and fourth of the fpurious ribs, reckoning from below upwards. But fince it appears from anatomy (*Albini hiftor. mufculorum hominis*) that the vault of the diaphragm afcends higher in the right fide of the thorax, for this reafon, when the perforation is made on the right fide, it fhould be performed betwixt the third and fourth rib; but when on the left fide, betwixt the fecond and third of the fpurious ribs.

Hippocrates takes notice, that if the matter or water be all of a fudden difcharged from a

patient

patient who has an empiema or dropfy of the thorax, it kills him; therefore fome would not have all the fluid extracted at once, but at feveral times. Now in an empiema or dropfy of the thorax, the lungs have lain a long time macerating in the matter, or in the extravafated ferum flowing all around, fo that upon evacuating it all, at one and the fame time, the lungs might have their weakened veffels burft by the fudden dilatation of them with blood, whence fudden death. What renders this operation the more eafily practicable, is the compreffure of the lungs by the extravafated humours, and the depreffure of the diaphragm by their weight, by which means thofe two organs are not eafily injured upon perforating the pleura.

The diaphragm is looked upon as a double and digaftric mufcle, made up of two different portions, one large and fuperior called the great mufcle of the diaphragm, the other fmall and inferior, appearing like an appendix to the other, called the fmall or inferior mufcle of the diaphragm.

The great or principal mufcle is flefhy in its circumference, and tendinous and aponeurotic in the middle, to this part the ancients gave the name of the nervous center of the diaphragm.

phragm. Senac has demonftrated, that the center or tendinous part of the diaphragm does not defcend in infpiration, the pericardium including the heart being attached thereto; for the pofition and motion of the heart would be difturbed fince the pericardium adheres with its broadeft fide to the tendinous part of the diaphragm. And that this part of the diaphragm does not defcend, he alfo proves, from its ftructure and connection. Lieutaud, in his Effais Anatomiques, afferts, that the contrary is eafily obferved in opening living animals. His words are, " l'on a dit que le centre tendineux " ne defcendoit point, à caufe de fes attaches " au mediaftin; mais il eft aifé d'obferver le " contraire dans l'ouverture des animaux vi- " vans."

This mufcle has a radiated flefhy circumference, the fibres of which it is made up, being fixed by one extremity to the edge of the middle aponeurofis, and by the other to all the bafis of the cavity of the thorax, being inferted by digitations in the lower parts of the appendix of the fternum, of the loweft true ribs, of all the falfe ribs, and in the neighbouring vertebræ.

We have therefore three kinds of infertions, one fternal, twelve coftal, fix on each fide, and two vertebral.

The

The fmall or inferior mufcle of the dia-phragm is thicker than the other, but of much lefs extent. It is fituated along the forefide of the bodies of the laft vertebræ of the back, and feveral of thofe of the loins.

The oval opening of this inferior mufcle gives paffage to the extremity of the æfopha-gus, and the aorta lies in the interftice be-tween the two crura.

In the middle aponeurofis of the great muf-cle, is a round opening which tranfmits the trunk of the lower vena cava. The circumfe-rence of this opening is formed by an inter-texture of tendinous fibres, and is confequently incapable either of dilatation or contraction, by the action of the diaphragm.

We find therefore three confiderable open-ings in the diaphragm, one round and tendi-nous for the paffage of the vena cava, one oval and flefhy for the extremity of the æfophagus, and one forked, partly flefhy and partly tendi-nous, for the aorta.

The ufes of the diaphragm are to affift in re-fpiration, in which it is a principal inftrument, defcending towards the abdomen in infpiration, and rifing upwards into the cavity of the tho-

rax

rax in expiration: we are alfo to look upon this mufcle as a power acting confiderably upon the ftomach, and abdominal vifcera employed in digeftion, for the diaphragm being depreffed at every infpiration, all the contents of the abdomen are thereby compreffed, and again in expiration they are repreffed by the abdominal mufcles.

From recollecting the various parts the diaphragm lies near to, and others with which it is immediately connected, it will not appear wonderful that an inflammation of it is fometimes taken for a diforder in fome of the other parts. From thence alfo it will appear, that various fymptoms may arife, according as different parts of the diaphragm become the feat of the difeafe. We fee for certain that it grows to the pericardium, tranfmits or gives a paffage to the æfophagus, aorta, vena cava, &c. lies clofely incumbent about the liver, fpleen, kidneys, and the reft—from whence an inflammation, fwelling, or hard fcirrhous tumour in this, or that part of the diaphragm may excite various complaints by injuring one or the other organ that lies next to it.

Leewenhoeck is perfuaded that the diaphragm puts the whole abdomen into a continual motion, whereby the food in the fto-

mach

mach and inteftines is comminuted in fuch a manner as to be reduced into a fluid matter, fit to enter the abforbing veffels fo numerous in the cavities of the inteftines. He computes that, in a well conftituted body, there is 900 refpirations every hour, and fo often will the ftomach and inteftines be compreffed.

Leewenhoeck rather inclines to Dr. Jurin's opinion, viz. that the palpitation of the diaphragm is better grounded than that of the heart.

The very accurate Winflow obferves, that the liver in the human body is fo firmly attached by its ligaments, that it cannot eafily flip from one fide to the other; yet that it is not abfolutely fufpended by them, but is in part fuftained by the ftomach and inteftines, efpecially when they are full. Hence after long fafting, the liver defcending by its own weight, pulls down the diaphragm, and occafions an uneafy fenfation, which the fame anatomift thinks is unjuftly afcribed to the ftomach.

We have well attefted inftances of very fmall wounds received in this part, through which fome of the bowels of the lower belly have afcended into the thorax. For while the contents of the abdomen are preffed by its mufcles

and

and the diaphragm, they are forced through
the wound in the latter, which they dilate, fo
as to pafs into the cavity of the breaft, and then
by compreffing the lungs and difturbing the
action of the heart itfelf, death is fooner or
later brought on.

Thus Parey affirms, he faw a man who was
wounded in the tendinous part of the diaphragm,
which, though no larger than the breadth of
one's thumb, the ftomach was notwithftanding
forced through the wound into the cavity of the
thorax. " Diffecto ventre inferiore (fays he)
" ventriculum cùm non reperirem, rem mon-
" ftro fimilem arbitrabar. Sed tandem anxié
" perquirens, raptum ipfum in thoracem ani-
" madverti, etfi vulnus pollicem vix effet latum."

In another perfon, who had been wounded
about eight months, and who died after the fe-
vereft cholicky pains, the inteftinum colon was
found the greateft part of it within the cavity
of the thorax, though the wound in the diaphragm was no larger than to be capable of
receiving the end of one's little finger. " Mor-
" tui cadaver Jacobi Guillemeau chirurgi pe-
" ritiffimi manu diffectum eft, animadverfum-
" que magnam coli partem, flatu multo tur-
" gidam, ipfi per diaphragmatis vulnus in tho-
" racem irrupiffe, vulneris tamen amplitudo
" vix minimi digiti capax erat." *Paræus lib.*
9. *p.* 30.   D   A

A like example is also to be met with in Sennertus, of a person who stabbed himself, but was cured of the wound within two months after, and notwithstanding which he expired, with frequent vomitings, seven months after the wound was healed. Upon opening the body, the wound appeared to have penetrated through the diaphragm and lungs, but the whole stomach was forced up into the left cavity of the thorax, and the heart with the pericardium was thereby pressed into the right side; when the patient was alive and cured of his wound, he would often direct those about him to feel the palpitation of his heart, by applying their hand. *Sennertus lib.* ii. *cap.* 13. *pag.* 372.

These are some of those wonderful sort of cases, for discovering which we have no certain signs, and can seldom know them but by opening the body after death, so that the errors which arise from our prognosticks of such wounds are unavoidable; for who could assert or foretel that the viscera were thus displaced, or by what signs could any one discover the same?

One ought to be well acquainted by anatomy with the true situation and connection of the diaphragm, and great caution is necessary in determining whether or no a wound penetrates into the cavity of the thorax, for as this cavity, as has been observed, ascends much higher

before

before, than behind, where it defcends lower,
very great errors have been fometimes commit-
ted, in thinking a wound penetrated the thorax,
when it really entered the cavity of the abdomen.
Thus we read in Ruyfch *(obf. anat. chir. obf.* 65)
of an ignorant Surgeon, who being defirous to
perform the paracentefis of the thorax, fent for
him into confultation; but Ruyfch being indif-
pofed and unable to go, the Surgeon by himfelf perforated the thorax as he thought; but
foon after, a large number of hydatids forced
themfelves out through the wound, and the
Surgeon, being affrighted, ftopped the wound
with a tent, and had recource to Ruyfch, but
to no purpofe, for the unhappy woman died
foon after; and upon opening the body, nothing of water appeared in the thorax, but the
Surgeon, in perforating the abdomen inftead
of the thorax, had wounded the liver, which
in that part adhered to the peritonæum, and
feemed to have degenerated into hydatids,
which burft forth through the wound.

Haller alfo tells us, that he opened the body
of a man who died foon after the operation for
the empiema had been performed on him. In
the liver there was a purulent ulcer, but not
deep, fituated very near the furface of it and
the diaphragm. He does not mean he fays to
throw the leaft reflection on thofe gentlemen who

had

had the care of this patient. To determine at all times where the liver terminates, and where the lungs, requires no fmall knowledge in anatomy, feeing the diaphragm diverges backwards from its anterior border, where it is higheft, and defcends from the fourth or fifth rib, as low as the twelfth, and even lower. Yet he concludes, opening the breaft, in a difeafe of the liver, does no great honour to the operator. *See his Pathological obfervations.*

Having done with our account of this part, we fhould next pafs on to thofe bowels in the cavity of the thorax, but before we proceed to examine thefe particularly, it will be firft neceffary to take notice of that membrane which lines its whole internal furface, known by the name of the pleura.

The pleura is a denfe compact membrane, which adheres very clofely to the inner furface of the ribs, fternum and intercoftal mufcles, and to the convex fide of the diaphragm ; it refembles the peritonæum, and likewife in that it is made up of an inner true membranous lamina, and a cellular fubftance on the outfide.

The cellular portion goes quite round the inner furface of the thorax, but the membranous portion is difpofed in a different manner. Each fide of the thorax has its particular pleura, intirely diftinct from the other, and making as it were

were two great bladders, fituated laterally with
refpect to each other in the great cavity of the
breaft, in fuch a manner as to form a double
feptum or partition running between the vertebræ and the fternum, their other fides adhering to the ribs and diaphragm.

This particular duplicature, which is formed
by a conjunction of the two facculi of the
pleura, is termed mediaftinum; the two laminæ of which it is made up are clofely united
together near the fternum and vertebræ, but
in the middle and towards the lower part of
the forefide, they are feparated by the pericardium and heart—a little more backward they
are parted in a tubular form by the æfophagus,
to which they ferve as a covering; and in the
moft pofterior part, a triangular fpace is left
between the vertebræ and the two pleuræ from
above downward, which is filled chiefly by the
aorta.

Before the heart, from the pericardium to the
fternum, the two laminæ adhere very clofely,
and there the mediaftinum is tranfparent, except for a fmall fpace near the upper part,
where the thymus is fituated; fo that in this
place there is naturally no interftice or particular cavity. Where the contrary has been fhewn,
it has been owing entirely to the common method of raifing the fternum, as is plainly demonftrated by Bartholine in his treatife of the
diaphragm.   At

At the upper part of the mediaſtinum, we find ſituated an oblong glandular body, called the thymus, which is round on its upper part, and divided below into two or three lobes, of which that towards the left hand is the long-eſt. In the fœtus it is of a pretty large ſize, leſs in children, and very ſmall in aged per-ſons. The greateſt part of the thymus lies between the duplicature of the ſuperior and anterior portion of the mediaſtinum, and the great veſſels of the heart, from whence it reaches a little higher than the tops of the two pleuræ, ſo that ſome part of it is out of the cavity of the thorax, and in the fœtus and children it lies as much without the thorax as within it.

There are various conjectures about its uſe, but as they are not ſupported by any proofs, ſhall forbear to mention them. And what the real purpoſes are, which it does ſerve, we are as yet unacquainted.

I once ſaw a ſuppuration in this gland in a lad I opened, and obſerved that the matter had emptied itſelf into the trachea.

Bonetus relates a caſe, where it was ſo much enlarged as to preſs upon the trachea, and oc-caſion a difficulty of breathing.

Platerus

Platerus, in his obfervations, tells us that
he has found it of fo extraordinary a fize, as
by its compreffion to caufe fuffocation and
death.

The mediaftinum does not commonly terminate along the middle of the fternum, as has
been generally fuppofed. And which at any
time may eafily be proved before the thorax is
opened, by running a fharp inftrument thro'
the middle of the fternum; we fhall find there
will be the breadth of a finger between the
inftrument and the mediaftinum, provided the
fternum remain in its natural fituation, and the
cartilages of the ribs be cut at the diftance of
an inch from it on each fide.

From all this we fee, not only that the thorax is divided into two cavities, entirely feparated from each other by a middle feptum without any communication, but alfo that by the
obliquity of this partition, the right cavity is
greater than the left.

The cellular portion of the pleura connects
the membranous portion to the fternum, ribs,
and mufcles, and in a word, to whatever lies
next the convex fide of the membranous portions of the pleura. It likewife infinuates itfelf between the laminæ of the duplicature, of
which

which the mediaftinum is formed, and unites them together. It even penetrates the mufcles, and communicates with the cellular fubftance placed at the back of the peritonæum, fo that if an inflammation is fucceeded by a fuppuration here, the matter confined in the cellular fubftance, that lies between the plates of the mediaftinum, may procure to itfelf wonderful paffages (as it defcends by its own weight) betwixt the pleura and the bodies of the vertebræ, where it may evidently form incurable finuffes and fiftulæ, and make its appearance by a fwelling in fome very diftant part of the body.——Inftances of this kind are by no means uncommon.

And to what amazing diftances matter will fometimes make its way, we read an example of in that excellent furgeon and faithful writer La Motte, " Traité complet de Chirurgie, " Tom. I. page 357."——where he traced the matter of an abfcefs from the loins to the fole of the foot.

This always happens in the panniculus adipofus; it not appearing, from any credible obfervations, that the proper fubftance of the mufcles has been at any time pervaded by a finus or fiftula.

We

We are well affured that the cellular membrane extends almoft into every part of the
body, and acquires different denominations,
according to the different matter which it contains. In thofe parts of the body where this
membrane is thinneft, its cells, being empty of
oil or fat, efcapes the eye, and is termed fim-
ply cellulofa—We fhould confider, that not
only all the mufcles and tendons are invefted
with fuch a cellular membrane, but that even
every mufcular fibre, as far as the eye and patient hand of the moft acute and dexterous
anatomift has been able to penetrate, is alfo
invefted with the fame; almoft every veffel in the body runs in or through fuch a
cellular fubftance, which in part conftitutes
the fabric of the veffels and vifcera themfelves.——*For a beautiful and extenfive account*
of this membrane, fee Scobingerus de dignitate
telæ cellulofæ in corpore humano.

To return to the pleura——We meet with,
in the celebrated Dr. Haller, a remarkable cafe
of an hydrops faccatus pleuræ, or an incyfted
dropfy of this membrane: upon opening the
thorax, to the no fmall aftonifhment of the
perfons prefent, there were no lungs (he fays)
to be found, but only a bag full of a green watery liquor, which, upon a further enquiry, was
obferved to have been extravafated between the
intercoftal mufcles and the pleura, and that this

E   mem-

membrane, the ufe of which is to line the infide of the ribs, was feparated from them in
fuch a manner as to form a bag as large as the
whole cavity of the breaft. Hence the left
lobe of the lungs was fo much compreffed, as
to be rendered thinner than one's hand, and
the cavity in which it was lodged no larger
than a glove. This (he obferves) is a very
rare cafe, and fhews, that an incyfted dropfy
may be produced in the thorax, from water
collected in the cells of the pleura.

The fame ingenious gentleman, from fome
diffections of perfons dying of a pleurify, is of
opinion, that the pleura is much feldomer the
feat of the difeafe, than is fuppofed; in thofe
he opened, the pleura was perfectly found, the
feat of the abfcefs being in that part of the
lungs which is contiguous to the diaphragm.
This obfervation, he fays, he could confirm by
innumerable inftances of the fame kind. To
this he adds, that the fpitting in pleuretic patients, owing to the inflammation, is eafily underftood, if the feat of the difeafe is granted
to be in the lungs; feeing, by the inflammation,
part of the obftructing matter will make its
way through the relaxed, exhaling, or mucous
veffels into the bronchia. But how that matter can pafs from the pleura to the afpera arteria, while the lungs remain found, he leaves
to thofe who are of that opinion to explain.

And

And from the rarenefs of the cafe of an inflammation of the pleura proving mortal, or
producing that pus which is collected in the
empiema, he would advife thofe who have the
treatment of pleuretic people, to fufpect the
caufe of that difeafe to be rather an inflammation of the lungs than of the pleura, and to
have immediate recourfe to the moft powerful
remedies.—See his Pathological obfervations.

In fupport of this opinion of Dr. Haller's,
I fhall bring the teftimony of Peter Servius,
from Triller's valuable treatife de Pleuritide,
who, after opening three hundred perfons de-
ceafed of pleurifies at Rome, always found one
lobe of the lungs corrupted and diftended with
a putrid matter, but the pleura appeared not
in the leaft affected.

Mr. Chefelden likewife ever found this to be
the cafe, (he fays) in opening fuch who were
fuppofed to have died of pleurifies.

It is no novel opinion, that the lungs are
the parts affected in a pleurify—for Aurelianus
reckons up a number of antient phyficians,
who defended this fentiment.

But to the mediaftinum, the ufes of this partition are to cut off all communication between the two cavities, by which, if one lobe

E 2 of

of the lungs is ulcerated, the other may not be immediately affected; that matter, water, or any thing elfe contained in one part of the thorax might not affect the other, and that, in cafe of a wound in one fide of the thorax, refpiration might be kept up in the other, and the perfon not immediately fuffocated. Another general ufe it has, is, to fupport the heart with its pericardium as well as the diaphragm, by means of which, the ftomach, the liver, and the other vifcera of the abdomen attached thereto, cannot, in drawing down the diaphragm too forcibly towards the inferior parts, interrupt the action of the heart and refpiration.

It may here be queftioned, whether a latent abfcefs under the fternum cannot have a difcharge of its matter by an opening through the fternum, left the pericardium and heart fhould be corroded by the confined and putrid matter?

Examples are not wanting in writers too of great credit, of the trepan's having been fuccefsfully applied over this part, in order to give vent to a collection of matter or water underneath.

Galen gives a remarkable cafe of a lad, who received a blow upon his fternum, and being negli-

negligently treated, that part of the sternum appeared afterwards corrupted by a sphacelus. Galen, trusting to anatomy in which he was so well skilled, cut out the carious part of the sternum, and tells us, he was able to see the naked heart, whose pericardium was eaten through, under the carious part of the bone; yet this lad was cured in a short time.

We are assured, by the celebrated Van Swieten, that he has seen matter, after a suppuration in the thorax, making its way through the sternum, which shews such a method of cure is possible.

We may therefore reasonably conclude, that when there are evident signs demonstrating a latent empiema seated beneath the sternum, an opening into it may be very safely attempted ; notwithstanding the assurance given us by Dionis in his anatomy, that he saw the sternum unsuccesfully trepanned in a wounded gentleman, who afterwards expired: for it does not in that case appear that his death could be ascribed to the operation, but rather to his wound; and enough has been said to shew that people have survived an opening made through their breast-bone, and as a further confirmation, shall produce one more instance from Haller's Prælectiones Academicæ, where we are told, that a certain Divine at Amsterdam had

had the trepan fuccefsfully applied to this part, and who was by this means freed from a great quantity of matter, and the heart was thus left fo naked, that by applying a looking-glafs oppofite to the wound, he could fee his own heart beating, which he appeared to take fome pleafure in, he being a man of uncommon tafte, and the author of a celebrated trea-tife, denying the exiftence of devils.

❊❊❊❊❊❊❊❊❊❊❊❊❊❊❊

PRÆLECTIO SECUNDA.

HAVING gone through the defcription of the pleura with its reduplicated plates compofing the mediaftinum, I fhall next proceed to the examination of the heart, with its capfula the pericardium.

The pericardium is a membranous fack of a pretty firm texture, which immediately includes the heart, and is placed between the two laminæ of the mediaftinum.

The capacity of the pericardium is equal to the fize of the heart, and is not as large again, as fome anatomifts have imagined; and if it feems larger than neceffary to cover the

heart

heart when a body is opened, it is wholly owing to the heart's emptying itself at the moment of death, not only of the blood contained
in its ventricles, but also that of the coronary
arteries, by which means, its volume is prodi-
giously diminished—This is an obfervation of
Lieutaud's.

The pericardium is compofed of two membranes, and of a cellular web which joins them
together: the outward membrane is tendinous
and very compact, and the inward is thin and
polifhed, being moiftened on all fides by a watery vapour. It muft be obferved, that not
only this cavity, but the interftices in the abdomen, thorax, and all other cavities of the
living animal, are replenifhed with moift va-
pours difcharged by exhaling veffels—the exiftence of which, in the fuperficies of the heart
and its auricles, may be proved by a fimilar
tranfudation of water or fifh-glue injected into
the large arteries: with this moifture, the heart
will be lubricated, and that friction prevented,
which might inflame and occafion its adhefion
to the pericardium. But if thefe exhaling arteries difcharge more plentifully than the veins
abforb, this fluid will be accumulated to the
quantity in which we often find it. By this
means, when the abforbing power declines in
chronical diforders, there is frequently produced a dropfy of the pericardium, and in
which

which cafe it has been found diftended with feveral pounds of liquor.

We may be fenfible how neceffary this moif-ture is to lubricate and feparate the vifcera, and efpecially the heart, from the hiftories given us by Peyerus, in which the patients were trou-bled with the moft violent oppreffions and pal-pitations, becaufe for want of this moifture the heart was found dry, and adhering to the pericardium.

The tendinous part of this bag has nine a-pertures or holes, as well for the paffage of thofe veffels which enter into its cavity, as thofe which iffue out of it; that is, two for the venæ cavæ, four for the pulmonary veins, one for the trunk of the aorta, and two for the pul-monary arteries.

The connexions of the pericardium are, with the principal veffels, juft now mentioned; it is alfo adherent, in a great part of its extent, to the aponeurotic portion of the diaphragm, fo as to be infeparable without laceration; by which means, it fuftains that broad mufcle with the feveral vifcera conneéted pendulous to it, fo that the diaphragm cannot by their weight be drawn down too low in the abdomen in our ereét pofture.

The

It has been faid, that this capfule has fometimes been found wanting.———Columbus tells
us that he opened a ftudent who died of a fit
in the univerfity of Rome, and that in this
fubject he could find no pericardium, but it
feems more probable, that from a preceding
inflammation, it had united itfelf fo clofely with
the heart as to have deceived him: certain it is,
that the coalition of this membrane is much
more probable, than the entire want of it.

Ruyfch kept by him the heart of a man
who laboured under a continual fever, with an
intolerable pain about the fore-part of his
breaft; but the outer furface of this heart was
altogether rough or unequal, from the pericardium being grown to its furface.

Lower produces a like inftance of its adhering every where fo clofely to the heart, that
he found it difficult to feparate it with his fingers.

It is probable, inflammations of the mediaftinum and pericardium happen oftener than
is commonly believed; for the caufes producing them may very powerfully act upon thefe
parts, namely, the cold air, or the drinking of
large draughts of cold liquors by perfons much
heated; but the pericardium is feated betwixt
the lungs, by which it is almoft every way furrounded, fo that the cold air infpired is, by the
F   dilated

dilated lungs, applied to the contiguous mediaftinum and pericardium, and the æfophagus,
tranfmitting cold drinks, go along betwixt the
two receding plates of the mediaftinum, and
paffes the diaphragm behind the pericardium;
the principal figns of this malady are fuch
caufes already mentioned, and a great heat felt
in the midft of the thorax; to which add a
great difturbance of the pulfe, and fainting
fits, when the inflammation has fpread to the
pericardium.—If a fuppuration fhall be formed
in the pericardium, it may penetrate the cavity
of that bag, and lie round the furface of the
heart.----Columbus (*de re anatomica, lib.* xv.
p. 267.) found the heart every way furrounded
by an abfcefs, by which it was almoft con-
fumed.

An inflammation of the mediaftinum is very
fairly confirmed by the obfervations of Avenzoar, who writes that himfelf laboured under
this malady—Upon his firft diforder, which
happened in a journey, he felt a pain in that
part, which increafed with a cough; he found
his pulfe very hard and his fever very acute.
The fourth night, he took away a pint of
blood; his fymptoms were but little relieved;
but in the night, and while afleep, the bandage
of the arm came off; upon waking, he found
the bed fwimming with blood, and foon after
recovered. The fymptoms in this difeafe are
generally,

generally, he fays, a continual fucceffive cough, a tenfive pain lengthways, a difficulty in breathing, an acute fever, great thirft, and a hard unequal pulfe.

This phyfician not only takes notice of an abfcefs in the mediaftinum, but in the pericardium likewife.

Salius Diverfus, who has withgood judgment given us an account of feveral diftempers, overlooked by the generality of writers, defcribes this diforder in a different chapter by itfelf, and fays, it had been taken notice of by no practical author before him. His defcription of the fymptoms, which follow upon an inflammation here, is very exact and particular; and becaufe the cafe is one pretty much out of the way, tho' without difpute, fuch as does often occur in practice, and may be eafily difcerned, if well attended to, I fhall juft give a fketch of what he obferves, (which indeed anfwers to what I have recited from Avenzoar) from the learned Dr. Friend's hiftory of phyfic—There is (fays he) an acute fever, inquietude, thirft, breathing thick and quick, great heat in the thorax, little pain except at the fternum, a cough always with it, and a hard pulfe. When the pericardium was inflamed too, there was a more intenfe heat, and a frequent fyncope; in one word, all the fymptoms worfe. And for a proof of what he afferts, he gives the cafe of

F 2   one

one, who died on the ninth day after some fits
of the syncope: where, upon dissection, there
appeared an inflammation of the intersepient
membranes, as he calls them, and some part
of the pericardium. And this distemper, I don't
question, happens oftener than our practitioners
commonly are aware of. He confesses that he
gave a more diligent attention to all these cir-
cumstances, because, being then young and
compleating his studies under very eminent
professors, he had seen a man of quality la-
bouring under the above distemper, which had
all the symptoms before enumerated, and who,
beyond all expectation, expired, when every
thing seemed to change for the better; but he
was several times troubled with fainting fits
before his death. But as the physicians were
here doubtful of the malady, and his friends
suspected poison had been given him, they de-
sired an enquiry to be made after the cause.
Upon opening the thorax, an inflammatory
swelling of considerable bulk was found in the
mediastinum; and an inflammation had in part
seized upon the pericardium. Nor was there
any other apparent cause of death found in the
body of the deceased.

Columbus takes notice of collections of
matter in the mediastinum, and which he and
Barbette order to be taken out by trepan-
ning the sternum.

As

As a further and more convincing proof of
what has been remarked, a gentleman, juftly
efteemed (fays Dr. Friend) for his long experience and found judgment in every thing re-
lating to furgery, informed him, that abfceffes
of the mediaftinum particularly happened in
venereal diftempers, and that, in fuch cafes,
he has frequently ufed the trepan with great
fuccefs. We may from hence be fatisfied, how
little ground there is for that hint of Parey,
where he feems to think this operation a ridi-
culous attempt.

Since the treatife of Peter Salius is rarely to
be found, you may read the chapter of Schen-
kius that contains the whole, " de inflamma-
" tione membranarum interfepientium et pe-
" ricardii—de tabe ex affectu pericardii—
" de tumoribus diaphragmatis."

Of the HEART.

The heart is a mufcular body, fituated in
the cavity of the thorax, being placed nearly
in a tranfverfe or horizontal pofture, with its
bafis in the right fide of the thorax, and its
apex in the left, while its broadeft and flat
fide, from the bafis to the apex lies inclined
and fupported on the diaphragm, to the tendinous or middle part of which it is firmly connected by the vena cava, and right venous

finus

finus below; and above, in the thorax, it is
connected within the duplicature of the mediaftinum, and lodged betwixt the foft lobes of
the lungs; by all which means, it avoids too
great a preffure on any fide, and is moft com-
modioufly adapted to receive the blood from,
and propel it into all parts of the body.

This is the true pofition of the heart in the
human body.—The figures in many of our
modern anatomifts are erroneous in this refpect—But the figures of Vefalius, Euftachius,
and Ruyfch, fhew the heart in its natural
pofture. From what has been faid here, we
may refolve the queftion, why over-eating
caufes a palpitation of the heart; for fince the
heart is only feparated from the ftomach by
the diaphragm, when the ftomach is over diftended, it wlll force up the diaphragm, and
prefs upon the heart. Hence alfo we may
fee how the heart comes to be preffed up fo
high in the thorax of thofe who die of a
dropfy in the abdomen, and why it is forced
fo far down in the abdomen of thofe who die
of a dropfy in the thorax; becaufe the diaphragm, to which the heart is connected by
its pericardium, is forced either way by the
contained water.

At the bafis of the heart are fituated two
muscular bags, one towards the right ventricle,

cle, the other towards the left, and joined together by an inner feptum, much in the fame
manner with the ventricles, one of them being
called the right auricle, the other the left.
They are very uneven on the infide, but
fmoother on the outfide, and terminate in a
narrow flat indented edge, reprefenting in fome
meafure the ear of a dog; they open into the
orifices of each ventricle, called by the name
of the auricular orifices, and they are tendinous
at their opening, in the fame manner as the
ventricles.

The right auricle is larger than the left,
and it joins the right ventricle by a common
tendinous opening, juft taken notice of. It
has two other openings united into one, and
formed by two large veins which meet and
terminate there almoft in a direct line, called
vena cava fuperior and inferior.

In opening a dog, not long fince, at the re-
queft of a lady, I found the pericardium amaz-
ingly ftretched with blood, fo as to fill half
the fpace of the thorax, with an aperture at
the meeting of the two venæ cavæ, large e-
nough to admit the end of a finger. In Bonetus's Sepulchret. Anatomic. we meet with two
or three inftances of ruptures in this vein in
the human fubject.—*Tom.* 1. *p.* 881. *fect.* xi.
de morte repentina.—*Obf.* 1. *Mors fubita ob ef-*
fufum

fufum fanguinem in dextrum cordis finum et peri-
cardium, á ruptura venæ cavæ.—Obf. 2. Cui-
dam morte concidente, fodiendi labore fatigato,
vena cava prope cor difrupta, fanguine inundavit
vifcera.

The whole inner furface of the right auricle
is uneven, by reafon of a great number of
prominent lines which run acrofs the fides of
it. In the interftices between thefe lines, the
fides of the auricle are very thin and almoft
tranfparent. There is an obfervation in Dionis, where he found this auricle fo far dilated,
that it was equal to the head of a new-born
infant.

The left auricle is in the human body a kind
of mufcular bag, or refervoir of a pretty confiderable thicknefs, into which the four pulmonary veins open. The whole common cavity
of this auricle is fmaller in an adult fubject,
than that of the right.

The flefhy or mufcular fibres of which the
heart is made up, are difpofed in fo fingular a
manner, that their courfe will be much eafier
underftood by feeing than defcribing them.

The heart is divided by a feptum, which
runs between its edges into two cavities called
ventriculi, one of which is thick and folid, the
other

other thin and foft : this latter is termed the right or anterior ventricle, the other the left or pofterior ventricle.

Lower, in his treatife *de corde*, tells us, that in a perfon who died of a confumption, and was fubject to fainting fits in his life time, he found both the ventricles of the heart nearly clofed up, particularly the right, by a flefhy fubftance, fo as fcarce to leave room for the admiffion of a goofe-quill.

We had lately a melancholly inftance of a rupture, in the fide of the right ventricle; and which occafioned the death of our late King.

Each ventricle opens at the bafis by two orifices, one of which anfwers to the auricles, the other to the mouth of a large artery, there-fore one of them may be termed the auricu-lar orifice, the other the arterial orifice. The right ventricle opens into the right auricle, and into the trunk of the pulmonary artery, the left into the left auricle, and into the great trunk of the aorta.

At the edges of thefe orifices are found fe-veral moveable pelliculæ called valves, of which fome are turned inward towards the cavity of the ventricles called tricufpides, others are turned towards the great veffels called femilunares.

G   The

The inner furface of the ventricles is very uneven, many eminences and cavities being obfervable therein; the moft confiderable eminences are thick flefhy productions, called columnæ, to whofe extremities are faftened feveral tendinous chords, the other ends of which are joined to the valvulæ tricufpides.

The valves at the orifices of the ventricles are of two kinds, one kind allows the blood to enter the heart, and hinders it from going out the fame way, the other kind allows the blood to go out of the heart, but hinders it from returning. The valves of the firft kind terminate the auriculæ, and thofe of the fecond lie in the openings of the great arteries.

The tricufpidal valves of the right ventricle are fixed to its auricular orifice, and turned inward toward the cavity of the ventricle. They are three triangular productions, very fmooth and polifhed on that fide which is turned towards the auricle, and on the fide next the cavity of the ventricle, they have feveral membranous and tendinous expanfions.

The valves of the auricular orifice of the left ventricle are of the fame fhape and ftructure, but are only two in number, and from fome fmall refemblance to a mitre, they have been named mitrales. Thefe five valves are very thin, and faftened by feveral tendinous ropes to the flefhy columnæ of the ventricles.		The

The femilunar valves are fix in number, three belonging to each ventricle, fituated at the mouths of the great arteries, and may properly be called valvulæ arteriales. The great artery that goes out from the left ventricle is termed aorta ; as it goes out, it turns a little toward the right hand, and then bends obliquely backward to form what is called aorta defcendens.

It is at the arch or curvature of this great veffel, that we oftner meet with the true aneurifm than elfewhere ; though Dr. Haller, in his pathological obfervations, cites an inftance of one in the carotid artery, which reached from the fubclavian to the divifion of the two branches of the carotid, and the furgeon, mifled by a kind of undulation, was preparing to lay it open.

It appears from anatomy, that the arteries, efpecially the larger, have pretty thick coats, partly cellular and partly mufcular ; fo that when the ftrength of the fides of an artery is diminifhed by any caufe, the confequence will be, that it will be dilated moft in its injured part, fo as to change the natural figure of the veffel, by diftending its weakeft part into a facculus.

We find, upon examining the hiftories of this difeafe, that they have for the moft part

G 2   arofe

arofe from blows and contufions about the breaft, fome again from an overftretching of their coats in violent ftraining, others from ero-fion.—Examples, where they have fprung from each of thefe caufes, we meet with in Lancifius.

Ruyfch has an obfervation of one at the cur-vature of the aorta, which was fo large as to equal a common cufhion.

It is found that in procefs of time the blood in thefe tumours begins to corrupt and be-come fo acrimonious, as to corrode the adja-cent foft parts, and does not even fpare the compact bones. And Ruyfch obferves in his cafe, that almoft all the ribs and fternum of the patient were reduced nearly to nothing. —Sometimes thefe poor unhappy patients ex-pire hereby in a moment. We read of a foldier, in the Academ. des Sciences, who fuftained a large aneurifm of this kind for fome time, when a flux of blood fuddenly burft forth from his mouth, of which he expired in a minute. Upon opening the body it was found, that the aneurifm adhered to the trachea, into which it had an opening betwixt the fixth and feventh cartilage, by which the blood efcaped into the windpipe and out of the mouth.

Surgeons fhould be very careful to diftin-guifh well this kind of tumour from others, fince

fince we are taught, by many obfervations, that feveral, in other refpects fkilful men, have imprudently deftroyed the patient by opening them.—That it is an aneurifm, may be collected—from the fore-mentioned caufes having preceded, from the tumor being feated in a part, where we know from anatomy, there is fome large artery feated, but more efpecially when it has a manifeft pulfation, and if the tumor diminifhes by a flight preffure, and returns again when the preffure is removed.

As aneurifms in the internal parts of the body are inacceffible, all that can be done for the patient is, to abate the impetus of the blood's motion by a thin diet, and repeated bleeding, by which means, the diforder may be prevented from increafing as much as poffible, and the patient at the fame time be ordered to refrain from all commotions of body and mind.

The trunk of the artery, which goes out from the right ventricle, is called arteria pulmonaris.—This I fhall leave to the particular hiftory of the lungs.

Befides the great common veffels, the heart has veffels peculiar to itfelf, called the coronary arteries and veins, fo named from their crowning in fome meafure the bafis of the

heart,

heart, they go out from the beginning of the aorta, and fend numerous ramifications to the fubftance of the heart.

This organ, with the parts belonging to it, are the principal inftruments of the circulation of the blood.

The heart is made up of a fubftance capable of cóntraction and dilatation; when the flefhy fibres of the ventricles are contracted, the two cavities are leffened in an equal and direct manner, not by any contorfion or twifting, as fome have imagined; and from confidering its ftructure, we muft fee, that it tends to make an even and uniform cóntraction, more according to the breadth or thicknefs, than according to the length of the heart, becaufe the number of fibres, fituated tranfverfely, is much greater than the number of longitudinal fibres.

The flefhy fibres thus contracted do the office of fuckers, by preffing upon the blood contained in the ventricles, which blood, being thus forced towards the bafis of the heart, preffes the tricufpidal valves againft each other, opens the femilunares, and rufhes through the arteries and their ramifications, as through fo many elaftic tubes.

The

The blood thus pufhed on by the contraction of the ventricles, and afterwards preffed
by the elaftic arteries, enters the capillary veffels, and is from thence forced to return by
the veins to the auricles, which, like porches
or antichambers, receive and lodge the blood
returned by the veins, during the time of a
new contraction.

This contraction is called the fyftole of the
heart.

The contraction or fyftole, ceafing immediately by the relaxation of the flefhy fibres, and
in that time, the auricles which contain the
venal blood being contracted, force the blood
through the tricufpidal valves into the ventricles, the fides of which are thereby dilated,
and their cavities enlarged.

This dilatation is termed the diaftole of the
heart.

In this manner does the heart, by the alter-
nate fyftole and diaftole of its ventricles and
auricles, pufh the blood through the arteries
to all parts of the body. Therefore the authors of the hypothefis, which makes the right
ventricle to contract itfelf before the left, have
been deceived, as Haller has clearly evinced,
by fome experiments you will find in a late
ingenious

ingenious treatife of his ' Sur le Mouvement
' du fang.'

Wounds, which penetrate the cavities of this
bowel, or by which any of the large blood
veffels iffuing out of it are opened, are confidered as abfolutely mortal.

It is to be remarked in wounds of the right
ventricle of the heart, that the lungs continue
to act, and by their dilatation, give an eafy
paffage to the blood to enter into them from
that ventricle; hence therefore there will not
be fo much blood expelled by the wound during the fyftole of the heart, becaufe of the
free paffage which it meets with into the lungs,
whence again fuch a wound will have the
greater opportunity to unite and heal.

But wounds of the left ventricle feem to be
much more dangerous, fince, if it be not totally perforated, the wound will of neceffity
be continually enlarging by the very ftrong
power with which the left ventricle contracts,
and which greatly exeeeds the force of the
right ventricle, in order to protrude its contained blood into the ftrongly refifting aorta,
fo as to dilate the fame and all its branches
throughout the whole body.

However,

However, we ought never to defpair even in the moft dangerous wounds. For there are fome obfervations which fhew, that men have often lived a confiderable time after wounds of the heart, efpecially when the right ventricle only has been perforated. Even fome obfervations teach us, that wounds of the heart are curable. *See Bartholine's Centur.* i. *Hiftor.* 77. ---*Schenckii Obfer. Med. Rar. p.* 275.---*Saviard and others.*

For while the patient continues only in a very weak and languid ftate, we may have feen in practice, I dare fay, that wounds have been healed, which no one would have thought poffible; and the likelieft method to fucceed is to keep the patient very quiet, and to avoid exciting the circulation by any ftimulus, efpecially thofe called cordials, which fhould be carefully avoided; the life of the patient may poffibly be preferved, and the wound healed.

For nobody would believe with how fmall a quantity and motion of the blood a perfon may live, who is not acquainted with the inftances given us by practical writers in the cafes of wounds, and in the mifcarriages of women.

H   It

It is to be remarked in wounds of the right ventricle of the heart, that the lungs continue to act, and by their dilatation give an eafy paffage to the blood to enter into them from that ventricle; hence therefore there will not be fo much blood expelled by the wound during the fyftole of the heart, becaufe of the free paffage which it meets with into the lungs, whence again fuch a wound will have the greater opportunity to unite and heal.

But wounds of the left ventricle feem to be much more dangerous; fince if it be not totally perforated, the wound will of neceffity be continually enlarging by the very ftrong power with which the left ventricle contracts, and which greatly exceeds the force of the right ventricle, in order to protrude its contained blood into the ftrongly refifting aorta, fo as to dilate the fame and all its branches throughout the whole body.

There is a diforder that pretty often occurs, though not much remarked, nor well defcribed, which is, *an aneurifm of the heart*, or a preternatural enlargement of its cavities; for while the heart's force exceeds the refiftance of the arteries, it continues of the fame dimenfions, but when the refiftance of thefe laft from

being

being grown rigid and boney exceeds the force of the heart, its cavities then enlarge.

Practical anatomy furnifhes us with many obfervations, teaching that the heart is thus frequently diftended: we meet with one in the philofophical tranfactions, where the left ventricle was found three times larger than the right.-----Fernelius, (in his Pathologia *lib.* v. *cap.* 12.) gives an account of a very uncommon and furprizing cafe of this kind ; where he fays, the frequent concuffions of the heart were fo violent and ftrong, as not only to luxate, but even to break fome of the adjoining ribs.

Marchettis in his anatomy, tells us, that he found a heart fo big, as to poffefs the whole thorax, and the ventricles of a prodigious extent, chiefly the right, that a natural fized heart might be contained in it.

But among the rarer caufes of this preternatural diftenfion of its ventricles, we may reckon the air, which has been fometimes found in the cavities of the heart, diftending them immenfely.

In a woman, who died fuddenly, Ruyfch found the heart of a ftupendous magnitude, from the air, with which it was full, contain-

ing

ing scarce any blood, as appeared from enter-
ing the point of a knife into it, the heart
suddenly subsiding, as if one had punctured a
bladder full of air.

Morbid dissections convince us, that in the
heart have been found, inflammations—suppu-
rations—erosions.

I shall add to that of our late King an instance of this sort, from the Academ. des Sciences—Where, upon opening the body of
the Duke of Brunswick, the heart was found
eroded by ulcers, and the right ventricle appeared burst, from such an ulceration and
erosion.

Morand, in the same papers, furnishes us
with another more surprizing instance----Who,
upon searching after the cause of sudden death
in the body of a nobleman, there appeared,
upon slitting up the pericardium, a large mass
of congealed blood, and in the left ventricle,
a perforation, which was equal to eight lines
in length, and the fleshy substance of the heart
appeared so infirm, that the probe made its
way through, in every part, by its own weight.
——The two foregoing extraordinary cases we
meet with, in the Acad. des Sciences for the
year 1732, under the following title, *Sur*
Quelques

Quelques Accidens remarquables dans les Organes de la Circulation du Sang.

When the caufe of a gangrene in the extremities of old people lies in the weaknefs of the heart, fo as to be unable to propel the blood into the extremities, we may well defpair of a cure.

In the following hiftories, the powers of the heart feem to have been fo weak, as not to be able to difentangle the veffels, which were folded together by the flighteft preffure that could be.

Tulpius records a remarkable cafe of an old dotard, who had long ftruggled with weaknefs, and the heat of the parts fo far extinguifhed, that every the leaft preffure upon any part of the body was immediately followed with a gangrene; fo that in a fhort time every part about him was almoft mortified before the poor miferable wretch was actually dead.

The celebrated Van Swieten tells us, that he had feen himfelf a refembling cafe in a woman of 90, whofe extreme parts were not only mortified before fhe died, but alfo the cheek which lay on the pillow while fhe flept.

From

From polypous concretions, whether formed in the heart or in its greater veffels, arife many irregular and terrible fymptoms.

Sometimes they have been found lying loofe in the cavity of the heart----at other times they have been found adhering to the veffels themfelves, and to the columnæ and auricles of the heart; of which there is a very remarkable one, both hiftory and reprefentation, in Bartholine's Centuriæ.------Thofe who are curious to know more of this matter, may confult Malpighius's treatife of the polypus of the heart, who was the firft (and fince him Ruyfch) who threw any light upon this fubject, which before his time was entirely in the dark.

There is but little hope of curing a confirmed polypus. All that can be aimed at, is, to dilute the blood, and fo throw it into a ftate moft remote from concretion, *i. e.* to introduce by art that cachochymy which confifts in the blood's being too thin, to the end the polypus may not be increafed by the appofition of new matter, but by degrees be worn away by the conftant attrition of the blood, which is every moment paffing by it.

PRÆLECTIO

PRÆLECTIO TERTIA.

HAVING confidered the heart with its pericardium, we come next to the defcription of the lungs.

The bags of the pleura are exactly filled by the lungs, which are two large fpungy bodies filling the whole cavity of the thorax, one being feated in the right fide, the other in the left, parted by the mediaftinum and heart, and of a figure anfwering to that of the cavity which contains them ; that is, convex next the ribs, concave next the diaphragm, and irregularly flatted and deprefled next the mediaftinum and heart.

The right lung is larger than the left, anfwerable to that cavity of the breaft, and to the obliquity of the mediaftinum, and more frequently divided or half cut through into three diftinct lobes or portions; but the left lung is not fo often divided into three.

At the lower edge of the left lung there is an indented notch or finus, (and which Euftachius

ftachius has taken care to exprefs in his Tables) oppofite to the apex of the heart, which
is therefore never covered by that lung even
in the ftrongeft infpirations, and confequently
the apex of the heart may always ftrike againft
the ribs, the lungs not furrounding the heart
in the manner commonly taught.

The fubftance of the lungs is almoft all
fpungy, being made up of an infinite number of membranous cells, and of different
forts of veffels fpread among the cells, in innumerable ramifications.

This whole mafs is covered by a membrane
reflected from each pleura. Betwixt the lungs
and pleura is found a watery or ferous vapour, of a coagulable nature, like that of
the pericardium, which vapour tranfudes from
the furface of the lungs and pleura, (*fee the
celebrated Kaau, de perfpirat. dicta Hippocrati
per univerfum corpus anatomice illuftrata*) and is
fometimes accumulated fo as to form a dropfy
of the thorax.

The chief veffels of which the fubftance
of the lungs is compofed are the air veffels and
blood veffels; the air veffels make the chief
part, and are termed bronchia, and which are
branches or ramifications of a large canal,

partly

partly cartilaginous and partly membranous, called trachea or afpera arteria.

The trachea or afpera arteria is a tube or canal, extended from the mouth down to the lungs. The larynx or head of the trachea forms the protuberance in the upper and anterior part of the neck, called commonly pomum adami.

It is conftituted of five cartilages, viz. the cartilago thyroides, which is the anterior and largeft, cricoides the inferior, and bafis of the reft ; two arytenoides, the pofterior and fmalleft; and the epiglottis, which is above all the reft.

The ventricles or facculi, mentioned by the antients and reftored by Morgagni, (who has given an excellent defcription of them, and of the whole ftructure of the afpera arteria) lie under the glottis, and are formed by a continuation of the internal membrane of the larynx.

The larynx ferves particularly to admit and let out the matter of refpiration ; and the folidity of the pieces of which it is compofed hinders not only external objects, but alfo any hard thing which we fwallow, from difordering this paffage. The glottis being a

very narrow flit modifies the air which we
breathe, and as it is very eafily dilated and
contracted, it forms the different tones of the
voice, chiefly by means of the different mufcles inferted in the arytenoide cartilages.

The facility of varying and changing the
tone of the voice depends on the flexibility of
the cartilages of the larynx, and decreafes in
proportion as we advance in age, becaufe thefe
cartilages gradually harden and offify, though
not equally foon in all perfons.

The general ufe of the epiglottis is to cover
the glottis like a penthoufe, and thereby hin-
der any thing from falling into it, when we
eat or drink; and which is fometimes the cafe
when we laugh or talk at the time of fwal-
lowing.

Haller tells us, that upon tracing the caufe
of low voice, and at laft an entire lofs of
fpeech in a woman. He found that one half
of the epiglottis was covered with an ulcerous
tumor here and there eroded, and which he
imagined to be the caufe of the defect in her
voice, and was aftonifhed that fhe was not
fuffocated.

The anterior and convex fide of the larynx
is covered with the thyroide gland. The
learned

learned Morgagni has remarked, that this gland is reprefented double by moft authors, though in reality it is fingle, and refembles the moon in its increafe, the horns pointing upwards.

Euftachius has very accurately delineated this gland, in his Anatomical Tables, but it is there four times fmaller than natural.

With regard to its ufe—notwithftanding the excretory ducts have not been afcertained, yet it is probable, from its fituation and connexion with the trachea and æfophagus, that they pour a lubricating fluid into them, and it is not likely, that fo large a gland upon the trachea fhould be fo clofely connected in that place without fome ufe.

Tumors very often happen in this gland; but fuch a fwelling is not properly a bronchocele, (though fometimes fo mifcalled) but a ftruma.

In morbid bodies this gland has been found enlarged to an extraordinary bignefs, fo as to reach down almoft to the clavicles; and in fuch cafes they generally turn fcirrhous ; when it is very large, neither any inward medicine nor outward application can diffolve it. Neither would any prudent furgeon, I prefume, at-

I 2   tempt

tempt to extirpate fuch a large tumor, for fear
of an hæmorrhage, or wounding the recur-
rent nerves. And Petit, in his Edit. of Pal-
fin's anatomy, vol. II. chap. ix. p. 313, gives
us a fufficient caution, in telling the ftory of
a daring furgeon, whofe patient in fuch a cafe
expired under his hands, to the great fcandal
of the profeffion.

There are other forts of tumors feated in
the forepart of the neck, and particularly defcribed by Albucafis, and which defcription I
fhall tranfcribe from Dr. Friend.

In treating of a bronchocele, or a rupture in
the forepart of the neck, which he fays is
moft frequent in women, he is fuller than the
Greeks or Celfus; he makes two fpecies, one
like a tumor which contains fome grofs fubftance, the other like an aneurifm. But tho'
he is fo bold in ufing the knife, he advifes the
operation only in the former cafe; and even
not there neither unlefs the tumor be loofe,
and little, and enclofed in a cyftis.

Sometimes thefe excrefcencies are full of
water, and fometimes they have nothing in
them but air.

This is a very frequent diftemper in thofe
countries where they drink great quantities of
cold

cold water, especially where they do not cool their waters in snow, as in other warm climates; but pour ice into it, as the way is with the ordinary people who live upon the black mountains of Genoa and Piedmont. The matter of fact is as true, as that they themselves attribute it to the drinking this water; and from the nature of cold, it is not difficult perhaps to account for the effect. And that the coldness not only of the liquors, but of the climate itself in other countries, may produce the same effects, seems to be plain, from the observations we find in writers, that these swellings about the throat and head, are more frequent among the northern nations than the southern.

We have several instances of diseases which are peculiar to some particular country, and seldom known any where else. Thus in Europe the plica polonica is proper to the Poles--- the scurvy to the borderers upon the Baltic Sea----and the guttur tumidum to such as dwell below the Alps. Thus in Asia, the vena medinensis or dracunculus is peculiar to the Arabians. So in Africa, the eliphantiasis was always the peculiare malum Ægypti, as we learn from Pliny.

The

The following quotation from De la Faye's notes upon Dionis, *des Operations de Chirurgie*, p. 640, will ſhew that a true bronchocele takes place here ſometimes.

" Le Goetre, comme Dionis le remarque,
" n'eſt pas une hernie, parce qu'il n'eſt pas
" formé de parties déplacées. Mais il ſurvi-
" ent quelquefois à la gorge une veritable
" hernie qu'on peut appeller proprement bron-
" chocele ou hernie de la trachée artére; car
" elle eſt formée par le déplacement d'une
" partie de la membrane intérieure de ce con-
" duit. Cette membrane en ſe dilatant paſſe
" entre les anneaux cartilagineux de la tra-
" chée artére, et forme a la partie antérieure
" du col un tumeur molaſſe, ſans douleur, de
" meme couleur que la peau, et qui s'etend
" quand on retient ſon haleine. Cette eſpece
" de maladie dont M. Muys dans ſes obſerva ·
" tions, et Manget dans ſes notes ſur Bar-
" bette, font mention, eſt fort rare, et nuit
" beaucoup a la voix et à la reſpiration."

The aſpera arteria or windpipe, conſiſting of ſemicircular cartilaginous ſegments with their back parts membranous and connected together by ſtrong muſcular ligaments, gives a free ingreſs and egreſs to the air from the glottis which is always open, and lined with a ſmooth lubricated membrane, ſo that

it

it will expand circularly by the air, give way
to the æfophagus in deglutition, follow the
pofture on bending of the neck, and become
either elongated or contracted, as there may
be occafion. The whole membrane, which
conftitutes the back of the trachea, where the
circular cartilages are deficient, is befet with
fmall glands, which feparate an unctuous hu-
mour, difcharged by their ducts, into the cavity of the trachea.

This canal is lined on the infide by a fine
membrane fo extremely fenfible, that nature
has placed it as a guard to watch at the door
of our breath and life, that we might not be
fuffocated by any particles falling into the trachea; for all bodies irritate it but the pure
air; even a drop of clear water excites a con-
vulfive and troublefome cough, which does
not ceafe till the liquor is ejected. This membrane has feveral fmall ducts (as has been obferved) opening into it, which continually pour
out a mucilaginous fluid, and which is of the
greateft ufe in defending it from being injured
by any acrimonious particles that may float in
the air.——This mucus, when left long undif-
turbed, may become fo tough and folid as to
take the form of blood veffels ; which, as
Ruyfch juftly obferves, has made fome of the
moft diligent obfervators give us hiftories of
clufters of veffels of great bulk being at once
ejected

ejected by coughing, (Tulpius among others
has done this with two figures reprefenting,
as he thought, the veffels coughed up) whereas
in fact it was only matter infpiffated and
moulded in the lungs.

If the windpipe be injured even with a
large wound, and the air has a free paffage
into the lungs, that wound will not always
prove mortal ; there are cafes which inconteft-
ably prove this, in which people being weary
of their lives have laid violent hands on themfelves, or in which the throat has been cut by
robbers, and yet they have been cured.

There are many who believe all wounds of
the trachea mortal, and fome have declared as
much in their writings. But the true ftate of
the cafe is, that when any of the large adja-
cent blood veffels are wounded, it is from the
injury done to them that the patient dies.

In Heifter's obfervations, there is an in--
ftance of a gun-fhot wound of the trachea
cured, and where a piece of the trachea was
carried away. *(See ob.* 80.)—and of which,
fays he, as much as I can call to mind, I have
not met with an example of in other authors.

Van Swieten remembers a foldier who ufed
to beg his way, and make a fhow of a large
wound

wound or aperture in the windpipe, which he would ftop with a fpunge, and then he could fpeak very well, but by opening the hole he loft his voice. This accident arofe from part of the windpipe being torn off by a bullet in battle, fo that the lips of the wound could not be afterwards brought together without leaving a confiderable aperture, yet he furvived the accident many years.

This laft cafe brings to my remembrance a device I ufed in a wound of this part, which a gentlewoman under a difcontent of mind had inflicted upon herfelf with a razor, juft below the larynx, between the thyroide and cricoide cartilages, with a lofs of part of the fubftance of the latter; as there was here (befides bringing the divided parts as near together as poffible) another intention, that of contributing to reftore the lofs of fubftance—I recollected Marchetti's contrivance in a fiftula which penetrated the afpera arteria, and which I fhall give in his own words, " Multas fanavi " fiftulas colli, potiffimum vero qua laboravit " adolefcens quatuordecim annorum, in parte " ejufdem anteriori, infra laryngem, oborto " primum ex contufione tumore, cui a barbi- " tonfore fanato, fuperftes fuit fiftula inter " utrumque annulum afperæ arteriæ, cum læ- " fione utriufque cartilaginis, ex qua fpiritus ef- " flabatur." Having firft prepared him (he fays) " Deveni ad topica; et primo quidem quatuor

K.   in-

" integumenta dilatavi turunda ex fpongia,
" deinde *fcalpro abraforio partes ipfas carti-*
" *lagineas utrinque læfas accurate abrafi,* tum
" filamenta arida fupra eafdem applicui; poft
" modum unguento iridis, quod occalefcere
" cœperat, in parte carnofa attrivi. Tandem
" globulis ex filamentis, imbutis unguento ex
" tutia, cavitatem opplevi; atq; fenfim et fen-
" fim imminutis globulis, cerato ex diachalci-
" tide, cum filamentis aridis cicatrice fiftulam
" obduxi.—Vide Marchettis Obs. Medico.—
Chirurgicar. Ob. 37, p. 52.

In imitation of the above method I flightly
(every other day) fcratched the edges of the
wound in the trachea with a fcalpel, dreffing it
up with dry lint and a retentive bandage, going on thus for about three weeks, I had the
pleafure to fee the opening contract apace,
when my patient, in the abfence of her nurfe,
and undefirous of living, accomplifhed what
fhe had before mifcarried in, (the deftroying
herfelf) by giving herfelf a fecond wound, and
dividing the left carotid artery hereby depriv-
ing me of the expectation I reafonably had of
fucceeding in the cure.

The ingenious Dr. Muzell, author of fome
medical and chirurgical cafes, which occurred
to him in the Charité at Berlin, affures us,
that he had feen the good effects of fcarification

tion in much the fame manner, in many cafes, particularly ulcers of the velum palati, and fiftulæ in perinæo——Patience and a conftant repetition of the fcarification is required ; for it is difficult to touch all the parts with the lancet, and they do not heal, till they become raw and inflamed ; he further fays, that he once faw this method profecuted with fuccefs in a diforder of the nofe, where there was a confiderable aperture in the cartilaginous part of the left noftril, which, after a long continuance of the fcarification, was cured. He directs that the incifions fhould always be made as near as poffible to each other, and to be repeated twice a day till the callofities are gone, and equal granulations of flefh appear.

Upon the facility with which even fome of the moft complicated wounds of the trachea arteria have been cured, bronchotomy has, both by the antients and moderns, been principally eftablifhed.

Sometimes extraneous bodies are fo engaged in the æfophagus, that we can neither extract nor deprefs them. And this happens when the extraneous body is of a confiderable bulk, and compreffes the trachea arteria to fuch a degree, that the patient is in imminent danger of fuffocation.

Habicot

Habicot (Queftion Chirurgicale fur la Bron-
chotomie cap. 16.) in this extremity infifts
upon the operation being performed as it ought
to have been (fays he) on him, who on a fefti-
val day, fwallowed a fmall bone of a leg of
mutton, which remaining in his pharynx, fuffocated him in the prefence of the phyficians
and furgeons, who did not relieve him by this
method.

A boy fwallowed nine piftoles wrapt up in
a linnen cloth, in order to hide them from robbers, but this bundle being too big, ftopped
at the narroweft part of the pharynx. He was
almoft fuffocated from the preffure made on
the trachea; his neck and countenance was fo
inflated, that his acquaintance did not know
him. Habicot, feeing him in fuch danger,
and not able to diflodge it, performed bronchotomy upon him; this was no fooner done,
than the inflation and livid colour of the neck
and face difappeared. And he made the parcel
defcend into the ftomach by means of a whale-
bone probe. Eight or ten days after, the patient at different times, difcharged the nine
piftoles by ftool.—This operation may alfo be
of another ufe than to make the patient re-
fpire, and which Habicot does not mention,
and that is to open a paffage for the extrac-
tion of bodies which flip into the trachea, and
are engaged in it.

Haller

Haller tells us in his Pathological Obfervations, that he diffected a boy who had been
fuddenly choaked by a filberd, which had ftuck
below the glottis, under the inferior ligaments,
below the thyroide cartilage, immediately above
the orifice of the afpera arteria. This misfor-
tune, fays he, might probably have been prevented by bronchotomy, if, when the accident
happened, the afperia arteria had been immediately opened, and fo the fatal nut taken out
with a fpoon.

The following cafe, though not relievable by
fuch an operation, I mention only as a very
extraordinary inftance of fuffocation, from the
fame author—A girl of ten years of age, and
whofe only complaint was worms; in diffecting
her, he found the mouth and throat quite full,
two of which he alfo found in the afpera arteria, near the feat of the heart, and at the beginning of the lungs.

Our countryman Willis feems to have been
the firft, who in fuch cafes thought of per-
forming the operation of bronchotomy—Willis
being once oppofed in a cafe which required it,
the patient died, and he afterwards performed
it in the prefence of thofe who were againft
it, and very eafily extracted a long and triangular bone.

Heifter

Heifter tells us, that he performed it on a young man, by making a longitudinal incifion, upon the trachea, and then cut through four or five rings longitudinally, and took out a mufhroom.

We are alfo told, that Rauu opened the trachea, in order to extract a bean, which had flipped into the larynx.

The obfervations are fo many, which might be adduced to prove the fafenefs of the operation, (and if an operation fo confiderable has been performed with fuccefs in a fwelling and inflammation of this part, it muft fucceed much more furely, when it is performed on a found part, whofe functions are only interrupted by the prefence of an extraneous body) that we muft neceffarily impute the death of fuch patients, to the timidity of thofe called unto their affiftance.

Verdue affures us, that this operation was in his time fuccefsfully performed by a furgeon, who was fo dexterous, as to extract by it a fmall bone through the opening, after which, the wound of the trachea was foon cured. Without this fpeedy and bold operation, continues he, nothing but death could have been expected.

Let

Let this ferve us as a caution on fimilar occafions, and let us not be fo cowardly and ti-
morous, as to let a patient die without affiftance ; for in cafes of neceffity every thing is to
be rifked.

We now and then meet with wounds of the
throat, through which the food efcapes; we are
not for this reafon always to fuppofe, that the
æfophagus and trachea are wounded——Thefe
wounds paffing fometimes only between the
glottis and root of the tongue.

Young furgeons fhould know, that a confiderable and fudden emphyfema is an occur-
rence not unufual in penetrating wounds of the
trachea, and this happens when the wound is
very fmall, or that in the trachea is not oppofite to that of the integuments; owing to
which it is that the air infinuates itfelf into
the cellular membrane, not only of the neck,
but the head, breaft, belly, fcrotum, and even
the upper and lower extremities, an inftance of
which I lately faw.—I fhall fay more of this
fymptom under my reflections upon the lungs.

The bronchial glands are of various fizes,
fome larger, fome fmaller; they are of a blackifh
colour, and are connected by a cellular fubftance to the loweft part of the trachea, and
to the divifions of the bronchia, and are fup-
                                                    pofed

poſed to communicate by ſmall openings with
the cavity of the bronchia.

The bronchia have their origin from the
trachea, and are firſt divided into branches afterwards ſubdivided again, and into almoſt innumerable ramifications : they finally terminate
in thoſe ſmall cells or veſicles, which form the
greater part of the ſubſtance of the lungs.

In two or three bodies dying of peripneu-
monies, Dr. Haller found theſe cells loaded
with blood, that a great part of the lungs were
become quite black, and though the lungs naturally ſwim when put into water, theſe on
the contrary ſunk, wherefore, he concludes,
that it is not in the veſſels only that the blood
of perſons in a peripneumony is congeſted, but
that there is a true error loci ; and that the
blood exhales into the veſicles of the lungs,
inſtead of that ſubtle vapour with which they
are naturally moiſtened.

Hence we are able to derive a truer idea of
the nature of inflammations in general ; and
that the blood in ſuch caſes is not alone ſhut
up within the veſſels of the inflamed part, but
that there is an effuſion of it alſo into the cellular membrane——This new, or rather re-
vived theory of inflammations (for he is ſo
candid as to ſay, *Hæc eſt antiquiſſima inflamati-*
onis

tionis theoria, quam Galenus propofuit) he has beautifully confirmed by inftances deduced from fome particular cafes of furgery, and which you will meet with in the firft volume of his Elementa Phifiolog. and further illuftrated in his Pathological Obfervations.

All the bronchial cells are furrounded by a very fine reticular texture of the fmall extremities of arteries and veins, which communicate every way with each other.

The blood veffels of the lungs are of two kinds; one common, called the pulmonary arteries and veins; the other proper, called the bronchial artery and vein.

The pulmonary artery goes out from the right ventricle of the heart, and its trunk, having run almoft directly upwards as high as the curvature of the aorta, is divided into two lateral branches, one called the right, the other the left pulmonary artery.

The right artery paffes under the curvature of the aorta, and is confequently longer than the left. They both run to the lungs, and are difperfed through their whole fubftance by ramifications nearly like thofe of the bronchia.

L. The

The pulmonary veins, having been diftri-
buted through the lungs in the fame manner,
go out on each fide by two great branches
which open laterally into the refervoir or mufcular bag of the left auricle. Thefe three or-
ders of veffels, with their courfe and numerous divifions, will beft be underftood by feeing fome preparation of them.

Befides thefe capital blood veffels there are
two others, called the bronchial artery and
vein by Ruyfch, who defcribed and gave
them that denomination, becaufe creeping or
fpreading upon the bronchia, they are extended
even to their extremities.

The great veffels appear deftined for the
circulation of the blood through the lungs.
The bronchial veffels are fuppofed to ferve for
their nutrition. For it appears to be a conftant rule of nature, that the vifcera, which
change by their fabric the common humours
brought to them for the ufe of the whole
body, have ftill other arteries peculiar to themfelves, which bring the vital blood appointed
for their nutrition.

The lungs are the principal organs of refpiration—hereby it is that parts noxious and
redundant are exhaled from the blood, and
there-

therefore not improperly defined by some the pulmonary transpiration.

Respiration is also necessary to facilitate the passage of the blood through the lungs, by which means it has the cohesion of its parts broken, attenuated, and pounded as it were. And it is by this force, that the abdomen with all its viscera are continually compressed. By virtue of this, the stomach, intestines, gall-bladder, receptacle of the chyle, urinary bladder, intestinum rectum, and the uterus itself, discharge their contents; and by this action, the aliments are principally ground or dissolved, and the blood is urged through the sluggish vessels of the liver, spleen, and mesentery.

We should scarce believe that part of the vital viscera might be amputated when they are exposed by wounds, if it was not proved safe by experience.

We read in Tulpius of a man who received a large wound under his left breast, part of the lungs on that side protruded through the wound, to the breadth of three fingers, which being neglected, begun to mortify, a ligature was immediately made round it, and then cut off with a pair of scissars, to the quantity of about three ounces, in fourteen days the wound was healed, and the man furvived for fix years afterwards.

Hildanus likewife certifies, that part of the lungs being prolapfed through a wound of the thorax, it was afterwards cut off with a hot inftrument of fteel, the patient being afterwards cured.

Nay Celfus formerly boldly pronounced, *Si quid aut ex-jocinore, liene aut pulmone duntaxat extremo dependeat, id præcidatur.*

Ruyfch tells us, (ob. 53.) that a fervant being wounded in the bottom of the fore-part of the thorax into its cavity, a furgeon was called, who feeing fomething thruft out of the wound, immediately made a ftrong ligature about the whole part that was expelled, thinking it to be a portion of the omentum, and immediately fent for him into confultation; but upon hearing him fay that the wound was not inflicted in the abdomen, but into the cavity of the thorax, and that the part thruft out, which he had tied with the ligature, was a portion of the lungs, he ftood like one aftonifhed. Things being thus ftated, Ruyfch advifed him to leave the wound as it was, till he found that part of the lungs intercepted by the ligature mortified, hoping that

it

it would be then feparated, and the remainder of the lungs healed in the wound itfelf, which fo happily fucceeded, that the patient was cured in a little time after.

There is a fymptom which is fometimes an attendant on penetrating wounds of the breaft and lungs, and of fo extraordinary a nature, that it will be inexcufable not to take notice of it. It is the emphyfema, or that amazing inflation which arifes from the air veffels of the lungs, being injured by the wound, fo as to depofit their injured air into the cavity of the thorax. It appears from medical obferva-tions, that the air having entered the cellular membrane may pervade almoft all parts of the body, and produce wonderful tumors, efpeci-ally as thefe cells are every where open, and form one continuous cavity throughout the whole body, fo that none of the cellular fa-bric is excepted from this communication, nay even the vitreous body itfelf of the eye has received the flatus of an emphyfema. We meet with (in the Academ. des Sciences) the hiftory of a man who was wounded into the thorax, and who, before his death, had his whole body furprizingly fwelled with one of thefe kind of tumors, excepting the foles of his feet and the palms of his hands; upon the thorax the tumor was eleven inches thick, and upon the abdomen nine, the eyes in this

fubject

subject were in a great meafure thruft out of
their orbits, from the cellular membrane being
diftended with a great quantity of air.

There is another cafe of this kind in the
fame work, of a fatal emphyfema, from a frac-
ture of the ribs, the fkin remaining entire.

Wherever fuch inflations are found, the
curative indication directs to difcharge the elaftic matter from the cellular membrane which it
diftends.—Parey gives us a fair inftance of the
fuccefs of fcarification in a cafe of this nature.
For while the miferable patient was given over
by every one, a fkilful furgeon (he fays) boldly
perforated the fkin with a great many very
deep fcarifications, and thereby difcharged the
included air, fo that the patient was reftored
to his health, and in a manner fnatched from
death.

The like tumors arife from a putrefaction of
the extravafated humours, fince it is evident
from experience, that putrefaction will produce or extricate the elaftic matter which lay
concealed in bodies, and which, if it is not
real air, has at leaft the fame elaftic power,
by which it will expand greatly by heat.

The air dilating itfelf in the blood after
death occafions various motions. Haller fays,
that

that he has feen the blood in the heart of a young man filled with air, to refolve itfelf into froth and to difcharge itfelf from the opening he made in the heart. He adds, that he has alfo feen the blood difcharge itfelf from the mouth of a very fine woman who died in child-bed, fo as to fill the fhroud.

There are many cafes of this nature; and perhaps this is the only caufe of the bleeding in dead people, which the fuperftition of the ancients hath regarded as an index of divine vengeance, which difcovered the criminal by this bleeding his prefence had created in the body of the perfon he had deprived of life.

To return to our reflexions upon the lungs —Should a large abfcefs in this part break fuddenly, fo that the matter cannot be dif-charged flowly, and in a fmall quantity at a time, by coughing, but vents itfelf at once in a violent flood into the windpipe, fo as to fill it, there will be no paffage left for the admiffion of the air; whence fudden death.— I faw two unhappy inftances of this not long fince.

Sometimes the abfcefs burfting, the matter falls into the cavity of the breaft, and this with a fatal event, unlefs there is a paffage
made

made for its difcharge by a timely aperture into the thorax.

Boerhaave faw, he tells us, in the body of one who expired of a fuppuration in the thorax, that the lungs on one fide were converted into a fack-full of matter, of fuch a prodigious bulk, that it not only difplaced the heart from its fituation, and compreffed the lungs into a fmall compafs, but likewife thruft down the diaphragm, and made it protuberate into the abdomen.

Haller likewife faw a moft terrible diforder of this part, in a body which he opened—The lungs on the left fide were not to be found, but inftead of them, a large quantity of matter; the afpera arteria (which if himfelf had not feen, he fhould fcarce have believed) and the larger arteries and veins opened with wide orifices into the cavity of the thorax, as if they had been cut through, fo that it was very hard to difcover what it was that prevented the efflux of the blood.

As the lungs from an inflammation often adhere to the contiguous parts, the abfcefs may break outwardly; which would prove a good effect, though from a bad caufe.

We

We are taught by morbid diffections, that the matter has eat through the diaphragm, and even into the ftomach itfelf; I have feen where it has got from the lungs through the diaphragm and into the liver.

Though the principal feats of calculi are the urinary or gall bladders, yet hiftory fhews us, that fcarce any part of the human body are excepted from ftony concretions, not even that organ we have now been defcribing.

The great Boerhaave relates, that the famous Botanift Vaillant fpit up four hundred ftones from his lungs, and that he has feen afthmas of the worft kind, where calculi have been brought up from the windpipe, when the refpiration would be free till more calculi were formed.

Kerckringius (in his Spicelegium Anatom.) has a cafe of this kind to which he has fubjoined the figure of the calculi adhering to the bronchia.

Le Dran tells us, that he knew two inftances, where they were often difcharged by expectoration.

There ftill remains another interefting remark to make upon this part. Every one has

M   heard

heard of the experiment upon the lungs made to difcover whether an infant has died before or after the birth.

It is the common opinion, that to be convinced of the truth in this cafe, we ought to put a piece of the infant's lungs into the water, and if it fwims, it is a proof that it has breathed, and confequently been alive.

Though this experiment appears convincing to abfolve or condemn perfons accufed of deftroying their children, yet it is proved, by many facts, that it is not fo infallible as it is imagined.

The lungs of a child, dead before its birth, fometimes fwim in water. This happens when, immediately after its birth, we blow into its mouth, as fometimes midwives do, when they doubt whether the child is really dead.

We meet with an example (in the Norimberg Tranfactions) by a phyfician; where, he fays, the child was certainly born dead, whofe body was not diffected, till it was confiderably putrified ; its veffels were full of air, and veficles, diftended with it, were feen on the lungs; pieces of the lungs fwam when put into water.

Hence

Hence we may see how great the uncertainty of this experiment is, and how cautious we ought to be, when we examine the bodies of infants, in order to make a report to a court of judicature: upon this head, *see Bohnius, de renunciatione vulnerum, differt. de infanticidio, pag.* 171.

FINIS.